Topical Conference: Sensors for Sustainability

Held at the 2024 AIChE Annual Meeting

San Diego, California, USA
27-31 October 2024

ISBN: 979-8-3313-1673-0

Printed from e-media with permission by:

Curran Associates, Inc.
57 Morehouse Lane
Red Hook, NY 12571

Some format issues inherent in the e-media version may also appear in this print version.

Printed with permission by Curran Associates, Inc. (2025)

For permission requests, please contact AIChE
at the address below.

AIChE
120 Wall Street, FL 23
New York, NY 10005-4020

Phone: (800) 242-4363
Fax: (203) 775-5177

www.aiche.org

Additional copies of this publication are available from:

Curran Associates, Inc.
57 Morehouse Lane
Red Hook, NY 12571 USA
Phone: 845-758-0400
Fax: 845-758-2633
Email: curran@proceedings.com
Web: www.proceedings.com

TABLE OF CONTENTS

FIELD-DEPLOYABLE SENSORS

GENERAL POSTER SESSION IN SENSORS

NEXT-GEN SENSORS

SENSORS AND MONITORING FOR HEALTH

STUDENT COMPETITION IN SENSORS (SPONSORED)

TOPICAL PLENARY: ENVIRONMENTAL ANALYTICAL TECHNOLOGY: SENSOR APPLICATIONS IN SUSTAINABILITY (INVITED TALKS)

618a Got Milk? Detecting Antibiotics with a Molecularly Imprinted Polymer Based Sensor

Isabel Chapa, Grace Dykstra, Yixin Liu

Sulfonamides, commonly used in veterinary medicine for the treatment and prevention of bacterial infections, can accumulate in dairy products from cattle, posing a public health concern. Several regulatory agencies around the world have established Maximum Residue Limits (MRLs), providing a demand for sensitive detection methods for these contaminants in milk, a complex mixture. In this study, a molecularly imprinted polymer (MIP) based electrochemical sensor was developed for sulfamethoxazole (SMX) detection. A systematic study of the electrochemical properties of SMX was conducted to identify an optimal detection solution that enhances the electroactivity of SMX within the MIP film. The monomer selected, pyrrole, was electropolymerized and overoxidized on a screen-printed electrode using cyclic voltammetry (CV) in the presence of a supporting electrolyte. An optimization of monomer and template concentrations, electropolymerization cycles, and pH was performed to achieve the desired selective analyte recognition. A comparison of the MIP and non-imprinted polymer (NIP) was done to validate the proposed sensor's characteristics. A nanomaterial layer under the MIP layer was used to boost the electroactive signal of SMX oxidizing, providing high sensitivity and a low limit of detection. Selectivity was investigated against other sulfonamides and other veterinary drugs that may be present in milk. Reusability, sensitivity, and reproducibility were also investigated. Lastly, the MIP-based SMX electrochemical sensor was integrated with a portable potentiostat to detect SMX in real milk samples. It holds great promise to detect SMX in dairy products in low-resources settings.

618b Multichannel Bioelectronic Sensing Using Engineered *Escherichia coli*

Xu Zhang, Marimikel Charrier, Caroline M. Ajo-Franklin

To advance environmental health and hazard detection, researchers have developed whole-cell bioelectronic sensors by engineering extracellular electron transfer (EET) to be dependent on an analyte. However, these sensors regulate a single electron transfer pathway as an electrochemical channel, limiting the sensing information to a single analyte. We have developed a multichannel sensor where different chemicals regulate distinct EET pathways within a single *Escherichia coli* cell. One channel utilizes the flavin synthesis pathway from *Bacillus subtilis* and is controlled by a cadmium-responsive promoter. A second channel, the Mtr pathway from *Shewanella oneidensis*, is controlled by an arsenite-responsive promoter through activation of cytochrome CymA expression. Further, we exploit the differing redox potentials of the two EET pathways to develop a redox-potential-dependent algorithm that efficiently converts electrical signals into 2-bit binary output. The algorithm enables the detection of heavy metal at EPA limits with a concentration-dependent response. When exploring our bioelectronic sensor in environmental water samples over the long term, a notable threefold decrease in current values was observed. However, the algorithm still allows for effective and accurate signal differentiation across analyte conditions. Thus, our multichannel bioelectronic sensor advances the field of simultaneous detection of different chemicals in a single cell, significantly expanding information transmission from EET and helping to safeguard human and environmental health.

618c H_2 Sensing with an Optical Fiber Sensor in the Subsurface H_2 Storage Conditions

Daejin Kim, Alexander J. Shumski, Ruishu Wright

Hydrogen can be stored in underground gas storage reservoirs in a large quantity for the high demand foreseeable in the near future. The concentration of hydrogen is subject to change due to biological reactions in the underground reservoirs and leakage through caprocks. Thus, monitoring hydrogen concentration in the subsurface gas storage deposits is vital to ensure the integrity and safety of the storage facilities. An optical fiber sensor is the most suitable sensing platform for the underground gas storage environments with the unique advantages of real-time and remote sensing without a risk of ignition and explosion. However, the optical fiber hydrogen sensor is not fully developed and validated for deployment under the harsh environment expected in underground gas deposits. In this study, the optical fiber hydrogen sensor was coated with a Pd/SiO2 layer and a filter layer, which enables H2 sensing in relevant subsurface storage environments. Specifically, the sensor has demonstrated successful sensing performance at high temperatures (~80 °C) and high pressures (~1,000 psi) under very humid conditions (~100% RH). In addition, the sensor was exposed to real subsurface microbial samples in the harsh environments described above to monitor microbially induced changes in hydrogen concentration. The sensor has shown stable H2 sensing responses in the replicated underground hydrogen storage conditions without deterioration or loss of H2 sensitivity. In addition, the biological subsurface sample with hydrogen gas resulted in higher transmission intensity change than the abiotic subsurface sample due to the possible hydrogen consumption with microbes. The successful demonstration of H2 sensing in relevant underground gas storage conditions would increase the readiness for deployment of optical fiber H2 sensors into the subsurface hydrogen storage reservoirs.

618d Novel Field Deployable Sensors to Monitor How Dynamic Hydrology Shapes Nutrient and Element Transformations in a Great Lakes Coastal Estuary: A Two-Year Study

Bukola Adesanmi, Sai Prasanna Chinthala, Terry Heard, Erin Eberhard, Timothy Morin, Lauren Kinsman-Costello, John Senko, Chelsea Monty-Bromer

The prediction of redox processes often relies on measured or assumed redox potential (Eh). However, process measurements frequently deviate from predictions based on the traditional thermodynamic "redox tower" paradigm. Poorly measured spatial and temporal soil heterogeneity has been shown to cause apparent departures from thermodynamic exclusion principles at bulk scales, impacting large-scale biogeochemical cycling unresolved in current ecosystem, regional, or global models.

To address this, we developed a 3D printed conductive carbon electrode array for zero resistance ammetry (ZRA) measurements in benthic sediments, enabling real-time detection of biogeochemical processes at various spatial scales. ZRA measures electrical current arising from microbially-induced redox disequilibrium. Prior to field deployment, we characterized the electrodes and conducted laboratory-scale experiments simulating redox conditions using Fe(III) (hydr)oxide-rich soils.

Refining our approach, we deployed a co-located ZRA and Eh multi-sensor system to detect electrochemical signals across and beneath the sediment-water interface at nested scales in the Old Woman Creek wetland. Concurrent data collection included dissolved oxygen dynamics, nutrient concentrations, greenhouse gas fluxes, and soil geochemistry. Future work aims to integrate microsite electrochemical data into ecosystem-scale models for improved representation of soil redox processes in spatially and temporally variable systems. Field data confirms the efficacy of the vertically stacked ZRA electrode in monitoring microbial processes at small scales.

618e Surfactant-Mediated Synthesis of Bismuth Oxychloride Nanoflowers for Electrochemical Detection of Heavy Metal Ions

Edward Fratto Jr., Cuilei Li, Zhengyang Yang, Mary Joens, Matthew Von Boecklin, Andrew Chapman, Zhiyong Gu

Contamination of municipal water with heavy metal ions poses a significant but often invisible risk to public health. While governmental agencies such as the Environmental Protection Agency (EPA) work diligently to set and enforce regulatory limits on heavy metal levels, their focus tends to be on localized hotspots in high-risk areas. Their sample analysis techniques are generally costly, inaccessible to the public, and require specialized training. Moreover, their oversight is limited to public infrastructure, leaving private water systems vulnerable to corrosion and contamination. While individuals can ship private water samples to local laboratories for analysis, practical monitoring of potential drinking water contamination requires more active and decentralized sensing methods to ensure reliability.

Recognizing the growing need for independent oversight in identifying and addressing toxic exposure, we proposed the use of high surface area bismuth oxychloride nanoflowers for detection of zinc and lead ions in drinking water. In consideration of the electrochemical methodology of sensing and their simple manufacturing procedure, they are an improved material for widespread use. Our method for the synthesis of the high surface area bismuth oxychloride nanoflowers involved the hydrolysis of bismuth chloride salt in an aqueous environment of sodium dodecyl sulfate. Generally, the reaction of bismuth chloride with water results in thin plate-like bismuth oxychloride structures primarily dominated by the 001 crystalline phase. We discovered that presence of surfactant sodium dodecyl sulfate was capable of crystallization behavior, encouraging the growth of 101 and 110 phases while reducing the 001 growth. The resulting structure was a uniform nanoflower material, as in the provided figure. This simple synthesis approach required no toxic materials for chemical reduction and proceeded immediately upon introduction of the salt into water, making it a highly scalable methodology.

In this work we proposed a mechanism for the formation of the uniform nanoflower structure, uniquely associated with the sodium dodecyl sulfate surfactant. We proposed that this growth is a result of the enhanced presence of the 110 and 101 crystalline phases, which undergo branching during crystallization in a uniform, concentric pattern. With additional surfactant addition, the restricted 001 growth encouraged further branching along additional angles from derivative phases such as 102, 103, 111, 112, etc., creating complex, high surface area flower structures. The as-synthesized nanoflowers were then suspended in Nafion solutions and deposited onto a glassy carbon electrode. Subsequent cyclic voltammetry analysis demonstrated the response to the analyte ions (Pb(II), Zn(II)), allowing for the detection of lead and zinc ions independently and simultaneously down to 1ppm. The implication of these results is the potential of bismuth-based structures for portable and accessible detection of heavy metal ions. Through parameter-focused structural optimization, we anticipate further improvements in sensitivity, paving the way for more effective monitoring of water quality and protection of public health.

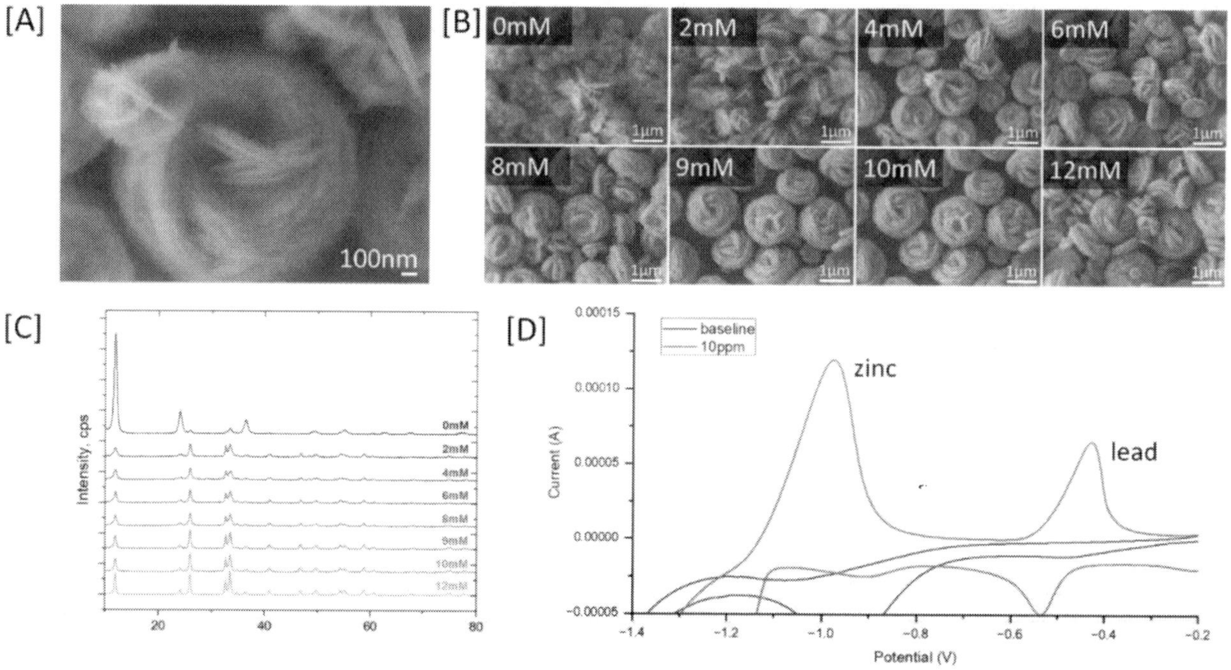

***Figure 1** – The presence of sodium dodecyl sulfate during the hydrolysis of bismuth chloride resulted in the formation of bismuth oxychloride nanoflowers with uniform morphology (a). The impact of the surfactant concentration on nanoflower growth and final flower formation was studied from a crystallographic perspective pairing observed morphology (b) with XRD characterization (c). These nanoflowers were capable of detecting zinc and lead (d) using electrochemical methods.*

618f A Novel System to Measure the Liquid/Vapor Fraction in a Geothermal Production Well in Real-Time

Zhonghua Zhan, Pengju Xing, Xiaolong Li, Marc Cremer, Andrew Chiodo, Joseph Moore, John McLennan

The liquid/vapor fraction, which is used to calculate production rate, is one of the most important parameters to monitor in real-time yet facing technological challenges. None of the state-of-the-art technologies can precisely measure the water/steam fraction in real time for an individual well at a relatively low cost. The geothermal system itself is hard to control precisely in terms of fluid flow rate, heat transfer, and homogeneity, which adds more complexity to the water/steam two-phase flow generated from a geothermal production well. Sensor technology to measure density, pressure, temperature, and velocity have been well advanced in the past a few decades. Two-phase flow homogenization devices have been investigated and applied in other chemical processing industries. If the advanced sensor technology and the two-phase flow homogenization technology can be combined, we shall be able to develop a water/steam fraction sensor and install it in a homogeneous non-steady two-phase flow environment to measure flow data of each individual production well in real-time. Here, Reaction Engineering International (REI) is teaming up with Energy & Geoscience Institute (EGI) at the University of Utah to develop a novel approach to accurately measure the water/steam fraction in a geothermal production well in real-time, where an accurate measurement of the density of the mixture is critical yet challenging due to the non-homogeneous and non-steady two-phase flow feature of the fluid coming out the production well. A bubble flow reactor is designed, machined, and assembled for this study. A homogenization device that produces a homogeneous water/steam two-phase flow is designed. The initial testing shows that the measurement from the newly developed sensor system is consistent with the designed conditions. Further improvements are still needed to increase the accuracy of measurement.

384a Cu_3n Sensors for Detecting the Air Pollutant NO_2 at Parts-per-Billion Levels

Adrien Baut, Andreas Güntner

In our current society, the quality of the air that we breath has become a major concern. Urbanization, industrial plants and emissions from transportation are all part of the main contributors to the decline of air quality. NO_2 is particularly problematic as it is a harmful secondary pollutant in presence with O_3 and directly affects human health. For that reason, its exposure limits are set in the parts-per-billion (ppb) levels by the World Health Organization. An effective NO_2 sensor would need to measure ppb concentration, be selective over many confounders, respond and recover fast, be robust to humidity and operate at low temperatures to limit the energy consumption. Chemoresistive metal oxides (MOx) are widely used for this task as they fulfil greatly some of those criteria (excellent sensitivity and low response and recovery times), however, they generally suffer from poor selectivity and high operating temperature. Here, we demonstrate the selective sensing of ppb-level NO_2 concentration in air at 50% relative humidity with highly porous Cu_3N films. The novel synthesis pathway [1] combines the aerosol deposition of a highly porous CuO film, followed by a dry nitridation to yield porous Cu_3N sensors. The sensor detects NO_2 down to a few ppb at 75 °C with excellent selectivity over aromatics, alkanes, alcohols, ketones and H_2 amongst others. As such sensors can be miniaturized and are low-cost, these can be applied as networks to provide distributed air quality assessment.

Reference:

1. Güntner, A.T.; Wied, M.; Pineau, N.J.; Pratsinis, S.E., *Sci.* **2020**, *7* (7), 1903390.

384b Aptamer-Enabled Solid-State Nanopore Sensors for Metal Ions Detection

Zehui Xia, Niloofar Zolfigol, Paul Nahass, Baikun Li, Yu Lei

Water quality monitoring is crucial during space missions, particularly on the International Space Station (ISS), where water resources are limited. The ISS requires a water monitoring system that is not only compact and user-friendly but also capable of sensitively monitoring drinking water quality to protect the crew's health and safety. A promising solution is the aptamer-enabled solid-state nanopore sensor. This sensor uses selective aptamers as molecular probes, significantly improving its ability to detect small molecules with high sensitivity.

In this study, we will first characterize various metal ion specific aptamers using Dynamic Light Scattering (DLS), Isothermal Titration Calorimetry (ITC), and other advanced techniques, and then demonstrate the remarkable sensitivity of our compact aptamer-enabled nanopore sensing system in identifying various water contaminants, such as silver, mercury, and lead ions. The integration of solid-state nanopores with aptamer-based signal enhancement enables the ultra-sensitive detection of these heavy metal ions. This study will showcase the aptamer-integrated solid-state nanopore sensor's capability for effective water quality monitoring in the challenging conditions of space. It holds the promise of ensuring the safety and well-being of astronauts.

384c Silver Nanoparticles Sensor Array for the Detection of Sars-Cov-2

Benjamin Lam

We report a silver nanoparticle (AgNP) sensor array to detect SARS-CoV-2 viral particles utilizing five peptide receptors to produce a colorimetric response. We show that the interaction of the virus particles with the peptides interferes with the shift in plasmonic absorbance and hinders the formation of snowflake-like fractal assemblies between the AgNPs and the bridging peptides. The limit of detection in water of the sensor array was 500,000 viral copies/mL. We demonstrate the capabilities of the sensor array to distinguish between SARS-CoV-2 (Beta, Delta, and Omicron variants) from the influenza virus at the 99% confidence level through linear discriminant analysis. The sensor array was also able to discriminate three out of five negative exhaled breath condensate samples from five positive SARS-CoV-2 EBC samples.

384d Real-Time Thickness Monitoring of Electropolymerized Molecularly Imprinted Polymers

Grace Dykstra, Yixin Liu

Molecularly imprinted polymer-based biosensors are synthetic alternatives to traditional enzyme and antibody-based sensing platforms. Electropolymerization, guided by cyclic voltammetry, is an easily controllable method for directly fabricating MIPs onto transducer surfaces, streamlining sensor design. However, the optimization of MIPs often relies on trial-and-error approaches, which lack fundamental understanding and increased experimental optimization time. This study investigates electropolymerization to achieve thickness-controlled MIP design. Leveraging insights from non-imprinted polymers, the impact of electropolymerization cycles on polymer thickness is explored. Polymer thickness is meticulously controlled and characterized using quartz crystal microbalance coupled with electrochemical measurements, aiming to aid in MIP design for enhanced biosensor performance.

441a Point of Need Biosensing of Protein Biomarkers from Complex Matrices

Yuqi Huang, Venkat Bhethanabotla

Biomarkers such as CEA, CA-19-9, CA-125, PSA, etc are currently quantified in diagnostic laboratories from body fluids to monitor progression, treatment and screening of various cancers, for which, point of need devices are not available. As with quantification of any protein from body fluids in a point of need (PON) format, the required low limit of detection for pg/ml determinations, interference form non-specifically (NSB) proteins present at many orders of magnitude larger concentrations, small sample size of a drop of blood vs venous draw, and long incubation times are challenges to be overcome by a PON platform. We present a PON sensor platform for quantification of biomarkers in body fluids using surface acoustic waves and nano-scale phenomena.

In our immunosensing approach, Rayleigh surface acoustic waves (SAWs) propagating in a delay-line SAW device are shown to be superior in replacing the current laboratory practice of extensive washing using buffer solutions to remove NSB interference. They are also shown to reduce incubation times by inducing acoustic streaming-based mixing of the sample, thus reducing assay times to minutes from hours/days. Guided SAWs with shear-horizontal polarization are shown to propagate in an orthogonal direction to the Rayleigh waves in certain piezoelectric substrates, which are effective in quantifying biomarkers in the liquid phase, leading to a compact device for simultaneous sensing, NSB removal and mixing. To lower the limits of detection to the pg/ml levels, we have explored nano-scale phenomena via gold nanoparticle (AuNP) based nanoprobes, and plasmonic enhancement of fluorescence to find the latter to be robust, albeit requiring additional optics to quantify fluorescence. We show that limits of detection are lowered by 2-3 orders of magnitude using these approaches. Multi-scale analysis of acoustic streaming, heat generation from streaming, and plasmonic enhancement of fluorescence have allowed us to both understand the fundamental phenomena and optimize the sensor device. We show that this orthogonal sensing device coupled with suitable electronics allows for constructing PON sensors for label-free, real-time sensing of biomarkers from complex fluids such as human plasma.

441b Development of Phosphate-Selective Sensors Based on Macrocyclic Ionophore-Doped Membranes

Monireh Dehabadi, Joshua Glazer, Andrew Schroth, Appala Raju Badireddy

Phosphate, widely used as a fertilizer, plays a crucial role in driving biogeochemical cycles and directly affects environmental quality, making accurate methods for phosphate determination crucial in fields such as environmental research and clinical chemistry. Released phosphate can accumulate in soils and sediments, causing pollution issues, eutrophication, and health problems. Due to the vital importance of phosphate, there is an urgent need for low-cost phosphate monitoring systems that do not require any chemicals, are not prone to interference from water turbidity, offer rapid response time, and simple fabrication.

Ion-selective electrodes provide the advantages of low cost, miniaturization, and portability, thus enabling the development of more suitable and practical methods for analyzing phosphate levels in the environment. In this study, two PVC-based ion-selective flexible sensors were developed by inkjet printing technique for sensing trace amounts of monohydrogen phosphate and dihydrogen phosphate ions in water.

Conductive nanoparticle ink with the concentration of 40 wt.% were printed on a flexible polyethylene terephthalate (PET) sheet with a thickness of 140 µm, forming reference and working electrodes (0.2 cm^2), respectively. The printing process was conducted using Fujifilm Dimatix 2850 drop-on-demand printer with a 10 pL cartridge. A nozzle temperature of 25 °C was used to generate stable droplets. The PET substrate temperature was set at 60 °C. Silver ink was printed on the electrode areas with a drop spacing of 10 µm for 5 layers, employing the water waveform, firing voltage from 23 to 30 V, jetting frequency of 5-30 kHz and cured at 140 °C for 30 min. The resulting electrode resistivity was 86 µΩ-cm.

Phosphate-selective membrane cocktails composed of high-molecular-weight polyvinyl chloride, PVC (33 wt.%) as the polymeric matrix, 2-nitrophenyl octyl ether, o-NPOE (63.5 and 62.6 wt.%) as the as plasticizing solvent mediator, ionophores (50 mmol/kg membrane), and hexadecyl trimethylammonium bromide (HTAB) as the anion exchanger (10 mmol/kg membrane) were prepared by dissolving in Tetrahydrofuran and investigated as $H_2PO_4^-$ and HPO_4^{2-} selective sensors. Ionophores were macrocyclic polyamide and polyamine compounds containing protonated nitrogen. The sensing layers were prepared by drop coating of each membrane cocktail onto the printed working electrode (WE). Potentiometric measurements were carried out at a constant temperature of 20 °C and pH of 7.4.

Potentiometric results showed that ionophore-based membrane electrodes exhibited a good sensitivity to aqueous solutions of the monohydrogen and dihydrogen orthophosphate species within the concentration range pC=2-5, with near-Nernstian response slopes of −28.3 and −50 mV per activity decade for ionophores I and II, respectively, with a response time of less than 60 s. Both sensors exhibited good reproducibility (SD<±3 mV). Selectivity coefficient values for orthophosphate ions were determined using the separate solution method and revealed negligible interference from anions, including sulfate, nitrate, and perchlorate.

Selectivity coefficients ($LogK^{pot}$)	Ionophore I	Ionophore II
NO_3^-	-2.7	-2.9
ClO_4^-	-3.8	-3.1
SO_4^{2-}	-2.3	-2.9

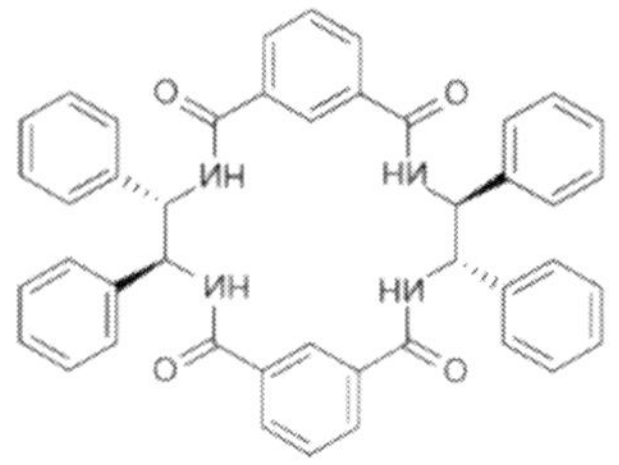

Ionophore I

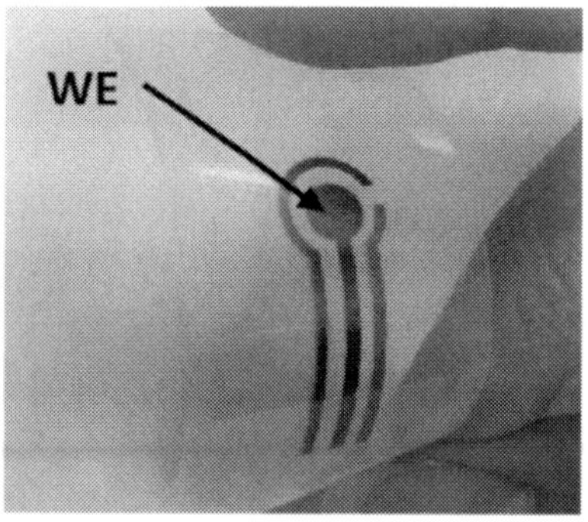

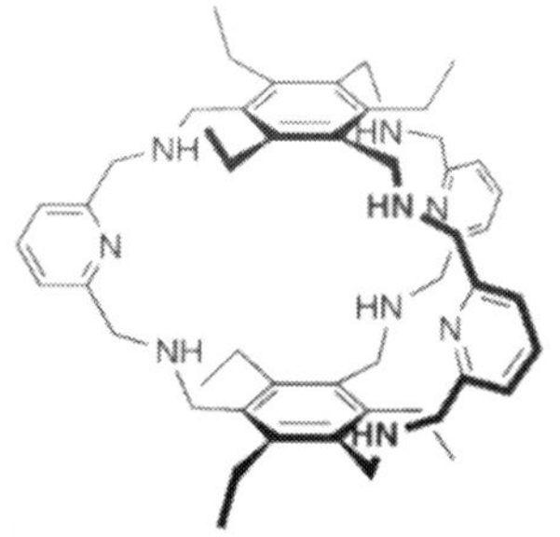

Ionophore II

441c Go with the Glow – Persistent Luminescence Nanosensors for pH, Oxygen, and Ion Quantification in Complex Samples

Tyler Sodia, Kevin Cash

Nanoparticle-based fluorescent sensors for tracking biological markers enable spatial and temporal tracking of analytes, revealing nutrient gradients, changes in signaling compounds, and other physiological phenomena. However, they suffer from interference from background fluorescence in optically complex samples – something present in most biological systems.

While there are many approaches to avoiding background fluorescence interference, one approach is to use persistent luminescence – "glow-in-the-dark". In this work, we synthesized near-infrared emitting $ZnGa_2O_4$:Cr^{3+} persistent luminescence nanoparticles (PLNPs) as a glow-in-the-dark source to incorporate into our established polymeric nanosensor platform. We combined these PLNPs with sensing components for a range of analytes (pH, oxygen, sodium, potassium, or calcium) to generate functional nanocomposites to sense these analytes in complex biological systems (including serum and yeast fermentation). In short, these nanosensors have the ability to glow with luminescence lifetimes orders of magnitude longer than biological background signals, enabling function even in optically complex samples.

441d Determination of a Broad Range of Volatile Organic Compounds in Environmental Air

Xiao-An Fu, Sujoy Halder, Zhenzhen Xie

Aldehydes including formaldehyde, acetaldehyde and acrolein and benzene, toluene, ethylbenzene, and xylene (BTEX), trichloroethylene (TCE) are toxic organic compounds in environmental air. Currently, there is no single method for sampling these compounds and no single method for analysis of these compounds. Current sampling and analytic methods for these compounds are not only time consuming, but also inducing more variation and errors. This work reports a novel microfabricated silicon preconcentrator device to combine physical adsorption of BTEX and derivatization of aldehydes in environmental air samples to enable analysis by a portable gas-chromatography-photo-ionization-detector (GC-PID). The captured compounds in the micropreconcentrator were eluted by solvent. GC-PID was used for analysis of the eluted liquid samples. The microprecocentrator was designed to have two separated compartments to accommodate two different sorbents. One compartment was loaded with physical sorbent like Carbopack X to capture BTEX, and the other was filled with silica gel particles coated with reagent O-(2,3,4,5,6-pentafluorobenzyl) hydroxylamine (PFBHA) to capture aldehydes via oximation. Both the single compartment devices with Carbopack X and PFBHA coated silica gel were first characterized for capture efficiencies of carbonyl compounds and BTEX. Then, the microprecocentrators with dual-compartments were tested for capture efficiencies of a mixture of BTEX and carbonyls in air. About 90% of recovery was obtained for BTEX, formaldehyde, acetaldehyde, and acetone by the micropreconcentrators. The limit of detection (LOD) was obtained in a range of 0.03 – 1.2 $\mu g/m^3$. The dual-chamber microfluidic devices were used to detect airborne BTEX and carbonyl compounds of environmental air samples in chemical industrial sites. The results indicate that the micropreconcentrators can be used for capture and analysis of a broad range of volatile organic compounds including aldehydes and BTEX by a portable GC-PID.

441f Temperature Sensor Application of Magnetic Phase Transition and Magnetic Moment Change

Hikaru Kiyomoto, Yuka Sakai, Yasuki Kansha

In recent years, many industries are showing great interest in sustainable and digitalised societies, with concepts such as cyber-physical systems (CPS). In doing so, efficiency data collection on a large scale and in real-time has become a crucial point of concern. From the initial data acquisition to the data storage and communication component of a CPS, maintaining a sensible efficiency and accuracy is of paramount importance for a sustainable society. One type of physical data that is crucial in various systems throughout society is temperature. Temperature is often one of the key indicators or parameters of any given physical system. Some of the conventional temperature-sensing methods include thermography and the use of thermistors or thermocouples. However, these methods have points of concerns, such as with the need for calibration, operation cost, and more. In this research, the authors propose a novel energy harvesting sensor with a possible application for a temperature data acquisition device. This can for instance be for monitoring body temperature, which is not only important for our society in terms of public health, but also a highly quantifiable measure for monitoring health. The proposed device utilises the magnetic properties of materials as the working principle in order to produce electricity from low-quality subject heat while simultaneously monitoring the temperature. When a magnetic material is subjected to an external magnetic field, the magnetic moments of the material align, thereby causing the material to magnetise. The degree of magnetisation, or magnetic susceptibility is characterised by Curie's law as being dependent on the material temperature. As the material approaches its Curie Temperature, a magnetic phase change occurs, causing a corresponding magnetic flux change. By incorporating electromagnetic induction with solenoids, the magnetic flux change of a known probe material could be measured as a voltage. Thus, this phenomenon can be applied for an energy harvesting temperature sensor in the near range of a material's Curie Temperature, as the voltage reading is representative of the magnetic flux change and thus the material temperature. With an appropriate material, the device has the potential to operate as a temperature sensor for various targets.

Past explorations by the authors have shown that the application of magnetic phase transition for a temperature sensor based on magnetisation change is especially promising in the phase transition temperature region around the Curie Temperature of the material. Therefore, it is necessary to identify and develop a probe candidate material that has the appropriate magnetic properties for the sensing subject. Here, an investigation on the probe material and its potential implications on the expected performance of the proposed sensor device. Specifically, a consideration of the magnetisation-temperature curve and electromagnetic induction magnitude is conducted on probe material candidates. Additionally, a theoretical approach to the formulation of a mathematical model for the sensing scheme has been conducted to quantitatively characterise the properties of the proposed device. The results of the study show the potential of the proposed device, with an appropriate probe material, as a temperature sensor for various sensing subjects.

441g PANI/MoS_2 based Room Temperature Gas Sensor for Sub-Ppm NH_3 detection: Experimental and DFT Insights

Aakanksha Jain, Ajithkumar MP, Siddhartha Panda

Background and Motivation:

The high levels of hazardous gas emissions have an adverse effect on both human health and the environment. Therefore, it is imperative to monitor harmful gases to protect both humans and the environment. This creates a substantial need for low-cost, dependable environmental monitoring and safety gas sensors.

Molybdenum disulfide (MoS_2), a form of 2D transition-metal dichalcogenide, has been extensively studied for chemical vapor detection because of its long interlayer distance (1.24 A°), high absorption coefficient, and large surface-to-volume ratio. MoS_2-based gas sensors, on the other hand, have a lower room-temperature sensing response and lack reversibility, which is most likely related to their low electric conductivity, ease of 2D nanosheet restacking, and significant oxygen adsorption in air. As a result, it encourages the fabrication of MoS_2-based sensors with superior sensing reversibility, chemical stability, and sensing responsiveness for use in practical detection applications. Meanwhile, conducting polymers provide room-temperature sensing and better selectivity, especially polyaniline (PANI). However, the environmental stability and sensing response of pristine PANI-based gas sensors is low, thus having a negative influence on their use as sensors. To improve sensing capabilities, several fabrication techniques have been put forth in the literature. These include creating p-n heterojunctions of PANI with metal oxides, carbon nanotubes, rGO, and noble metals, which would alter the band structure and best utilize the characteristics of each component in the composite. Few studies, nonetheless, have discussed the use of PANI in conjunction with two-dimensional (2D) materials.

In our work, the PANI/MoS_2 composite, screen-printed on PET substrate, is utilized as a flexible room-temperature-based sensor to investigate its behaviour with ammonia (NH_3) and various other gases with biological and environmental implications. The sensing capabilities are analysed experimentally and through theoretical studies, i.e., DFT studies. Further, two different synthesis routes have been explored to synthesize the composite to study the effect of the morphology of PANI/MoS_2 on sensing.

Methodology:

Different routes for the synthesis of PANI-MoS_2 composite.

1) *Route 1:* PANI-MoS_2 composite was synthesized using interfacial polymerization of aniline in the presence of bulk MoS_2. Further, to evaluate the influence of MoS_2 in sensing and find a suitable ratio that provides the best sensing performance, polymerization was carried out in the presence of varying concentrations of MoS_2.

2) *Route 2:* Multi-layered MoS_2 nanosheets were developed using hydrothermal treatment via precursors (ammonium molybdate and thiourea with a specific ratio reported elsewhere). Few layers MoS_2 nanosheets synthesized by hydrothermal methods were added to the polymer reaction system

from the initiation stage. After the pre-optimized duration of 12 hours, the polymerization was stopped, and through a rigorous purification process, the composite material was collected and stored under vacuum conditions until further used.

Once the composite formation was confirmed, conducting inks were formulated using α-terpinol. The chemiresistive-based sensors were fabricated by screen printing on PET sheets with specifically designed silver electrodes. The screen-printed devices were dried at 80 °C for four hours and stored in a vacuum desiccator until further use.

Sensing:

This study was carried out using an in-house gas sensing system where compressed dry air (CDA) is used as a carrier gas. A regulated amount of CDA is sent to the sensing chamber to remove any impurities in the chamber or previously adsorbed species. The analyte gas was transported to the mixing chamber in a particular ratio with CDA to achieve the necessary gas concentration. The mixture was fed to the sensing chamber and exposed to the sensor film. After a successful sensing investigation, an inert gas purge was performed to remove any remaining vapors from the sensing chamber. The relative change in the resistance of the sensor brought on by exposure to the target gas was used to measure the sensor response. All the measurements were conducted at a relative humidity (RH) of 10% and a temperature range between 25-30 °C, i.e., room temperature.

Results and Implications:

The sensor response of the composite markedly improved from that of the pristine PANI, showing an improvement of 11% for 10 ppm of NH_3. The reason for the superior sensing response of the composites over the pristine PANI is the synergistic effect of conducting polymer and 2D material; the EIS studies indicated that the p–n junction depletion layer formed at the interface of PANI and MoS_2 enhanced the sensing response. Additionally, the gas-sensing properties of the PANI/MoS_2 composites were reinforced by their abundance of reactive sites.

Using the PANI/MoS_2 device, the detection limit for NH_3 was found to be 122 ppb with a 2.9% response. There was no need for external heating because the sensing device could sense and recover at ambient temperature. One essential factor that impacts sensor responsiveness is humidity. The sensing response of the device was evaluated at different relative humidity levels, and it was discovered that the response was steady up to 70% relative humidity, but above that point, an increased response to NH_3 was seen. Such behavior might be because absorbed moisture or the production of conductive H_3O^+ further protonate PANI. Over nine months, the response values of the sensor fluctuate. However, this fluctuation is minimal, with a divergence of 2% demonstrating the consistency of the sensor over time. The composite synthesized via two routes exhibited structural and morphological diversity, demonstrated by all the characterization techniques. Morphology, however, had no appreciable effect in enhancing the gas-sensing response. This might be explained by the low content of MoS_2 in the composites, which lessened the impact of morphology on sensing. However, during sensing studies, it was observed that composites synthesized via route 2 showed better noise resistance and durability against environmental effects than the ones synthesized via route 1. Furthermore, the detection limit of NH_3 was in the sub-ppm range (122 ppb), but the limit of detection for other gases, including hydrogen sulfide, acetone, ethanol, carbon dioxide, and methanethiol, was in the ppm range (20 ppm-70 ppm) which is high compared to NH_3, signifying the sensitivity of our device towards NH_3. Added to that, to confirm sensing experiments and determine the rationale for sensing, a DFT analysis (Bader charge, charge density difference, density of states) was conducted, and one of the significant findings was that adding MoS_2 to PANI increased NH_3 adsorption while reducing the adsorption of other gases making this device more sensitive towards NH_3, which supported our sensing results. Furthermore, hydrogen sulfide, carbon dioxide, and dimethyl disulfide were shown to have physisorption, whereas NH_3, NO_2, acetone, ethanol, and methanethiol showed quasi-molecular interaction.

Further, the ability of our device to respond selectively to low quantities (sub-ppm range) of NH_3 suggests that the material has a significant potential for the non-invasive diagnosis of kidney diseases in humans as NH_3 in human breath exhaled by healthy individuals ranges between 20–688 ppb, but those with renal disorders have levels around 2000 ppb.

441h Multichannel Automated Temperature Stimulus of Elastin-like Polymers

Stanley Feeney, Jeffrey Halpern

Elastin-like polymers (ELPs) are proteins similar to naturally occurring elastin protein that have a tunable lower critical solution temperature (LCST). ELPs have been investigated for use in wound dressings, drug delivery, and biosensing. The stimulus response of ELP to changes in salt concentration and ligand binding have been characterized [1], as well as the response to temperature while in solution. However, the temperature-dependent behavior of surface-bound ELPs has not yet been fully characterized. This is in part due to the difficulty of efficient and accurate temperature control in analytical environments, especially in electrochemical techniques. A system has been built that allows for simultaneous analysis of 8 different electrodes in a temperature controlled electrochemical cell. We have been actively using Electrochemical Impedance Spectroscopy (EIS) to monitor the attachment of ELPs to electrodes as well as the stimulus behavior and stability of the resulting surfaces. We are also studying the effects of drift in EIS at different temperatures and solution conditions.

[1] M.A. Morales, W.A. Paiva, L. Marvin, E.R.M. Balog, J.M. Halpern, Electrochemical characterization of the stimuli-response of surface-immobilized elastin-like polymers, Soft Matter. 15 (2019) 9640–9646. https://doi.org/10.1039/c9sm01681c.

Support was provided by the National Science Foundation EPSCoR award (#2119237), BIO-SENS.

441i Responsive Complex Emulsions: All-Liquid Sensors for Detection and Classification of Environmental Contaminants

Suchol Savagatrup

Complex emulsions with responsive surfactants provide a unique sensing platform that leveraged tunable interactions at liquid-liquid interfaces. In this presentation, we report our progress toward the fabrication of distributed, real-time sensors by incorporating responsive complex emulsions with PDMS-based microfluidics. Specifically, we measure the dynamic interfacial behaviors of environmental contaminants (e.g., PFAS) in real-time and use these time-series data as chemical fingerprints to classify the identity and quantify the concentrations. We will discuss our effort toward applying a multiplexed platform for complex matrices of groundwater and mixtures of analytes. We anticipate that the incorporation of responsive all-liquid materials into sensing platforms will provide a cost-effective, easily tunable, and portable alternative to laboratory-based analytical methods for on-site sensing applications.

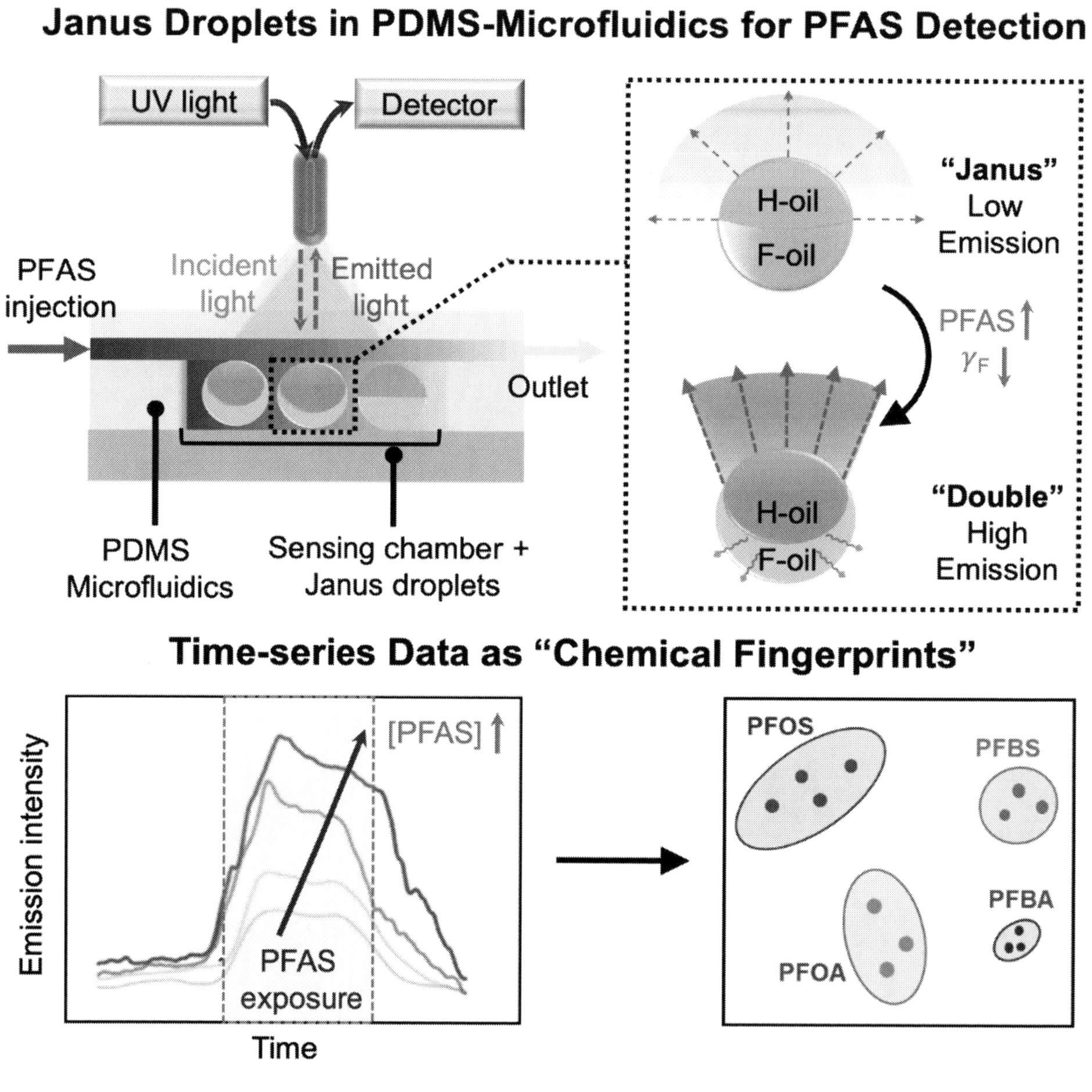

441j Assessing the Impact of Internal Flow Dynamics on Low-Cost Optical Particle Counters

Livingstone Quarshie, Douglas Webb, Matthew Jones, Randy Lewis

Environmental pollution represents a pressing global concern, prompting international efforts to mitigate the emission of harmful pollutants into the atmosphere. These pollutants emanate from a diverse array of sources, including construction activities, agriculture, power generation, transportation (particularly from automobiles), and the utilization of cookstoves for cooking. Among the primary pollutants of concern are Oxides of Nitrogen (NOx), Oxides of Sulfur (SOx), Particulate Matter (PM), and Ammonia. Notably, PM has garnered significant attention due to its adverse health impacts on the human populations upon inhalation [1].

PM is classified based on their aerodynamic diameters into three categories: fine particles ($PM_{2.5}$) with diameters less than or equal to 2.5μm, ultra-fine particles ($PM_{0.1}$) smaller than 0.1μm, and coarse particles (PM_{10}) with diameters less than 10μm. The health implications of PMs are size-dependent, with $PM_{2.5}$ posing a greater risk as they can penetrate deeper into the respiratory system, leading to conditions such as asthma, bronchitis, and mortality [2].

In the United States, monitoring instruments for PM include the Federal Reference Method (FRM) and the Federal Equivalent Method (FEM), which, however, report PM concentrations over extensive geographic areas [4]. To facilitate localized, real-time measurements, mobile PM sensors have been developed [5]. Various methodologies, including gravimetric, optical, and microbalance techniques, are employed for measuring PMs. Among these, the optical method, particularly Optical Particle Counters (OPC), is most prevalent. OPCs operate by drawing particle-laden air through a system where particles scatter light in a scattering region, with the intensity of scattered light measured by a detector to indicate particle concentration [6]. The scattering region is the region in an OPC where a particle interacts with light to cause scattering. There are expensive (has mirror or lens) and low-cost (no mirror or lens) OPCs. Low-cost OPCs are readily available to purchase and use.

Unfortunately, the design and operational efficiency of OPCs raise several critical questions concerning their ability to accurately measure ambient particle concentrations and sizes [7]. Factors such as the concentration of particles drawn into the OPC compared to ambient levels, the geometry of the OPC, the impact of fan speed, and variations in concentration within the OPC are crucial considerations. Furthermore, determining the appropriate sampling time is essential for accurate measurements.

This work investigates the performance of OPCs through two approaches. Initially, a comprehensive Computational Fluid Dynamics (CFD) analysis of simple defined geometries, which mimic the tortuous path flow characteristic of commercially available OPCs, was conducted. This foundational analysis elucidated the impact of device geometry on variation in concentrations and subsequently on measurement accuracy. Leveraging on insights gained from the initial CFD study, the research was expanded to explore the performance of commercial low-cost OPCs. This examination focused on addressing critical design and operational questions previously highlighted. Through this inquiry, the research provided substantive evaluations and potential enhancements for the accuracy and reliability of low-cost OPCs in real-world environmental monitoring applications. By employing two CFD software packages for comparative analysis, the research also enhanced the understanding of OPCs' performance and identified potential improvements for their accuracy and reliability in environmental monitoring.

In the initial CFD study of the simply designed geometries, SolidWorks, a computer-aided design (CAD) software, was utilized to create two geometric models: a singular box chamber and a dual-box chamber. The dual-box chamber was employed to investigate the effects of geometric variations on OPC accuracy. The method described in this section focuses more on the use of Converge studio for the initial simple geometric studies.

The simple designed geometries were imported into Converge Studio CFD for simulation setup. The simulations employed a transient turbulence model (RANS-k-epsilon), ensuring detailed flow dynamics analysis. Injector specifications and orientations were defined for each geometry's inlet, alongside comprehensive boundary condition settings for inlets, outlets, and walls, adhering to a no-slip condition for the stationary walls.

The discrete phase model (DPM) in Converge Studio was used. Air served as the continuous phase, while carbon particles represented the discrete phase, and a Rosin-Rammler particle size distribution was used to inject particles with varying sizes. The particles injected were given a Sauter mean diameter of 2.5μm, targeting a concentration of 35μg/m³. Following case setup, input files were generated, and simulations were run. Tecplot360, a post processing software, was used to analyze the results from the simulations of simple geometries. To facilitate a detailed investigation of regional concentrations at various time intervals, a custom Python script was developed. This script strategically segmented each geometry into distinct regions, enabling precise calculations of regional concentrations over different simulation times. Plots of various regional concentrations versus time were generated for different time averaged concentrations. Particle size distributions plots were also generated, enabling a better understanding of different particles sizes in various regions at different times. The simulations in Ansys Fluent CFD also used the DPM model, similar parameters, and the same assumptions as used in Converge studio CFD. Two CFD packages were used to provide additional validation of the simulation results.

The findings from the simulations of the simple geometric models revealed notable variances in particle concentrations across the chamber's regions. While some regions exhibited significant deviations from the target concentration of 35μg/m³, other regions closely aligned with or distributed around this target level. Furthermore, the size distribution plot underscored disparities in particle size distribution within the regions.

Additionally, two commercial OPC designs were also modeled using CAD to capture all relevant geometrical components influencing measurement outcomes. The analysis between two commercially available OPCs delved into the disparities in concentration and particle size distributions across various regions in the geometry.

The insights derived from this study are expected to inform strategic considerations in selecting specific region(s) within an OPC for the scattering region and detector position, leading to improved measurement accuracy. Through these endeavors, the research aims to advance the development of more accurate and reliable OPCs, ultimately contributing to enhanced environmental monitoring capabilities and a deeper understanding of the role of PM in air quality and health.

References

1. Choi, Y.J., et al., *Particulate matter(10)-induced airway inflammation and fibrosis can be regulated by chitinase-1 suppression.* Respir Res, 2023. **24**(1): p. 85.
2. Cormier, S.A., et al., *Origin and health impacts of emissions of toxic by-products and fine*

particles from combustion and thermal treatment of hazardous wastes and materials. Environ Health Perspect, 2006. **114**(6): p. 810-7.
3. U.S Environmental Protection Agency(EPA). *2024 PM Nation Ambient Air Quality Standards Final overview presentation*. 2024; Available from: https://www.epa.gov/pm-pollution/national-ambient-air-quality-standards-naaqs-pm.
4. Christopher, S.A. and P. Gupta, *Satellite remote sensing of particulate matter air quality: the cloud-cover problem.* J Air Waste Manag Assoc, 2010. **60**(5): p. 596-602.
5. Steinle, S., et al., *Personal exposure monitoring of PM2.5 in indoor and outdoor microenvironments.* Sci Total Environ, 2015. **508**: p. 383-94.
6. Amaral, S., et al., *An Overview of Particulate Matter Measurement Instruments.* Atmosphere, 2015. **6**(9): p. 1327-1345.
7. Trejo, R.E.G., et al., *A Study on the Behavior of Different Low-Cost Particle Counter Sensors for PM-10 and PM-2.5 Suspended Air Particles*, in *Telematics and Computing*. 2022. p. 33-50.

295b Advancing Personalized Healthcare through Wearable Biosensors

Jihong Min, Wei Gao

The evolution of personalized healthcare is significantly propelled by advancements in wearable bioelectronic systems, which enable non-invasive, real-time monitoring of vital biomarkers through biofluids like sweat. These systems hold the promise of transforming healthcare by facilitating continuous health status assessments outside clinical settings, thereby aligning interventions with individual health needs. However, the practical realization of these technologies faces hurdles in autonomous biofluid access, the development of highly sensitive and selective sensors for a comprehensive biomarker spectrum, sustainable energy sources for operation, and the scalability of manufacturing processes (**Figure 1**)[1].

Electrochemical Detection of Essential Biomarkers.

Central to overcoming the challenges in personalized healthcare monitoring is the creation of novel sensors capable of detecting a wide range of biomarkers, including electrolytes, metabolites, amino acids, proteins, and hormones. The most mature sensor technologies in this domain, enzymatic sensors and ion-selective electrodes, are adept at detecting a handful of metabolites and electrolytes but are limited in their scope. Consequently, the monitoring of many key biomarkers necessitates the adoption of alternate or novel sensing technologies. This research propels the development of electrochemical sensors employing advanced techniques such as imprinted polymers (MIPs), aptamers, and antibodies to bridge this gap.

These sensors, integrated with scalable fabrication methods such as laser engraving, inkjet printing, and additive manufacturing, afford unprecedented sensitivity, selectivity, accessibility, and scalability. Specifically, the use of laser-engraved graphene (LEG) electrodes and microfluidics, patterned using a CO^2 laser cutter, exemplifies low-cost scalability while maintaining high performance. This approach not only enhances the feasibility of widespread application but also marks a significant advancement in sensor technology.

The utilization of molecularly imprinted polymers (MIPs) alongside antibody-based sensors marks a transformative advancement in biosensing[2]. MIPs, synthesized to mimic natural antibody functions, bind specific biomarkers with high fidelity, mirroring the selectivity and sensitivity of biological systems but without their inherent limitations. These 'artificial antibodies' were engineered through a process where functional monomers form a complex around a template molecule, which, once polymerized and the template removed, leaves behind a cavity perfectly shaped for the target biomarker. When integrated with laser-engraved graphene (LEG) electrodes, this approach facilitates the precise detection of a wide range of biomarkers directly from biofluids such as sweat. This combination not only leverages the conductivity and surface area advantages of LEG but also utilizes the selectivity of MIPs, enabling the monitoring of essential health indicators like amino acids and vitamins with unparalleled accuracy.

The antibody-based CRP sensor platform integrates laser-engraved graphene electrodes with microfluidics on a patch, designed for the real-time monitoring of the inflammatory biomarker C-reactive protein (CRP)[3]. Utilizing gold nanoparticles (AuNPs) conjugated with antibodies and redox-active tags, this system amplifies the electrochemical signal upon CRP binding, ensuring highly sensitive detection. The microfluidic design facilitates efficient sweat sampling and mixing directly on the skin, enabling the CRP antigen-antibody reaction to occur seamlessly within the device. Clinical studies conducted on patients have demonstrated the platform's efficacy, showcasing its potential for non-invasive chronic disease management by providing accurate, real-time inflammatory status monitoring through sweat analysis, thus heralding a significant advance in personalized healthcare monitoring.

Sustainable Powering of Wearables.

The autonomous operation of wearable health monitoring devices requires a reliable and continuous power source. Energy harvesting from the body, through biofuel cells (BFCs)[4] that convert metabolites like lactate into electricity, and triboelectric nanogenerators (TENGs)[5] that convert motion into electrical energy, presents a sustainable approach for powering wearable devices for fitness monitoring applications. However, these methods typically rely on exercise for sweat or power generation, limiting the applicability of wearable devices for continuous health monitoring, especially in sedentary or low-activity scenarios.

Addressing the limitations of body-powered systems, this research harnesses ambient light as an ubiquitous energy source. The development of flexible perovskite solar cells (FPSCs) offers a compelling solution, demonstrating high power conversion efficiency under both indoor and outdoor lighting conditions[6]. Distinguished from conventional silicon solar cells, FPSCs exhibit flexibility, robustness, and an optimal spectral response to indoor lighting, making them ideal for wearable applications. This energy-harvesting approach eliminates the dependency on physical activity for power generation, significantly enhancing the operational capability of wearable devices for continuous monitoring across various user activity levels.

The high power output of the FPSCs, coupled with a powerful low power electronic system, enables the platform to perform autonomous sweat extraction via iontophoresis and conduct all commonly used types of electrochemical measurements. This integration facilitates the interfacing and multiplexing of various sensors, allowing for comprehensive biomarker analysis. Moreover, data transmission is achieved wirelessly over Bluetooth Low Energy (BLE), ensuring seamless communication. By leveraging ambient light for power and enabling autonomous sweat extraction, this system markedly extends the wearable devices' operational capability for continuous monitoring, irrespective of the user's activity level. In a demonstrative application, biomarker levels (glucose, pH, sodium, skin temperature, sweat rate) were simultaneously monitored for over 12 hours using this battery-free device while a subject engaged in various daily activities, including exercise and desk work. This breakthrough underscores the potential of FPSC-powered wearable devices in realizing the vision of continuous, unobtrusive health monitoring, thereby advancing personalized healthcare through innovative technology.

By integrating novel wearable sensor technologies for metabolic and nutritional analysis with advanced energy harvesting methods for sustainable powering, this research lays the groundwork for the next generation of wearable health monitoring platforms. These platforms promise to enable the continuous, autonomous monitoring of a broad spectrum of health biomarkers, powered sustainably by our bodies and the environment.

References († indicates equal contributions)

1. Min, J.†, Tu, J.†, Xu, C†., Lukas, H†. *et al.* Skin-Interfaced Wearable Sweat Sensors for Precision Medicine. ***Rev. 123***, 5049–5138 (2023).
2. Wang, M.†, Yang, Y.†, Min, J.† *et al.* A wearable electrochemical biosensor for the monitoring of metabolites and nutrients. ***Biomed. Eng. 6***, 1225-1235 (2022).
3. Tu, J., Min, J. *et al.* A wireless patch for the monitoring of C-reactive protein in sweat, ***Biomed. Eng. 7,*** 1293–1306 (2023).
4. Yu, Y., Nassar, J., Xu, C., Min, J. *et al.* Biofuel-powered soft electronic skin with multiplexed and wireless sensing for human-machine interfaces. ***Robot. 5***, eaaz7946 (2020).
5. Song, Y.†, Min, J†, *et al.* Wireless battery-free wearable sweat sensor powered by human motion. ***Adv. 6,*** eaay9842 (2020).
6. Min, J.†, Demchyshyn, S.† *et al.* An autonomous wearable biosensor powered by a perovskite solar cell. ***Electron. 6***, 630-641 (2023).

Figure 1. Insights into Wearable Sweat Sensing Technologies for Customized Health Management

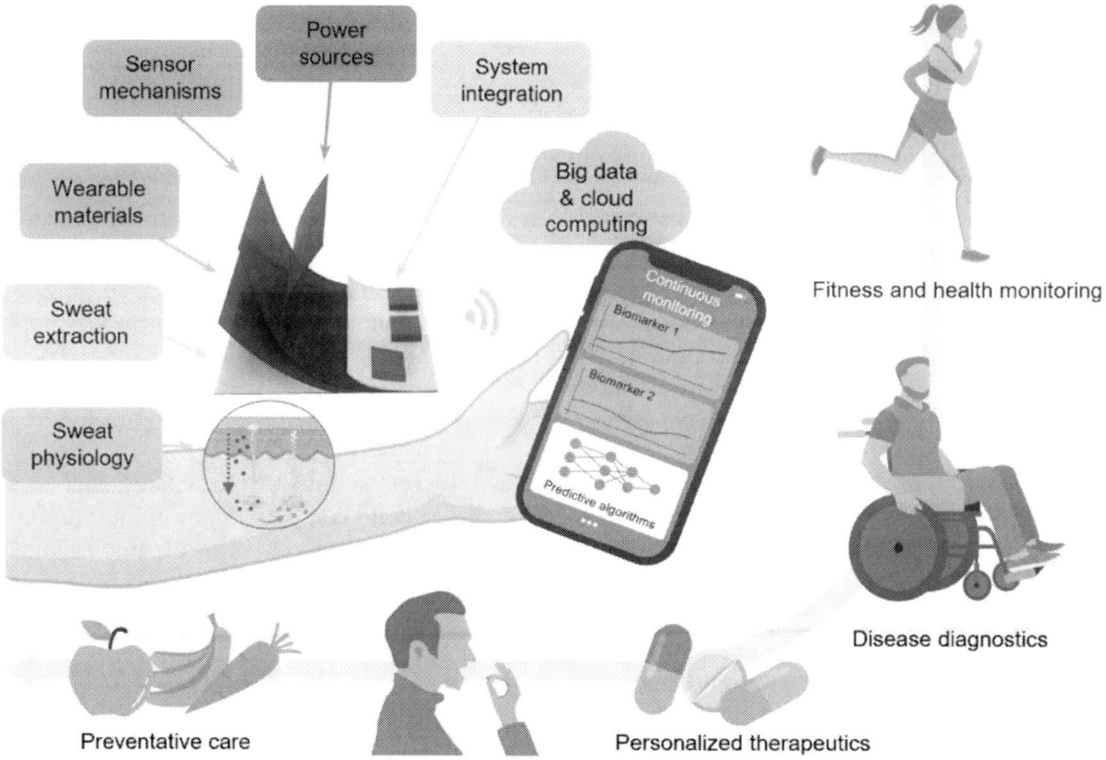

Power sources
Sensor mechanisms
System integration
Wearable materials
Big data & cloud computing
Sweat extraction
Sweat physiology
Continuous monitoring
Biomarker 1
Biomarker 2
Predictive algorithms
Fitness and health monitoring
Disease diagnostics
Preventative care
Personalized therapeutics

295c Amplifying Flutamide Sensing through the Synergetic Combination of Actinidia-Derived Carbon Particles and WS_2 Platelets

Rupa Kasturi Palanisamy, Yiran Luo, Carmel Breslin, Eithne Dempsey

The research into electrochemical sensors for detecting flutamide has reached a critical juncture in the realms of biomedical and environmental monitoring. A study was conducted where spherical carbon particles (CPs) were obtained from Actinidia fruit juice through a simple Hydrothermal method. These CPs exhibit a unique characteristic of appearing green under UV light and fluorescing with a maximum emission upon excitation with light at 350 nm. Despite having an average diameter of 500 nm (Spherical Shape), the presence of carbon dots decorating the CPs appears to be responsible for this unusual fluorescence. The CPs were then combined with tungsten disulfide (WS_2) using ultrasonication and centrifugal mixing, resulting in the formation of a CPs-WS_2 nanocomposite. The interaction between WS_2 and CPs showcased a clear synergetic effect, with their sizes and dimensions facilitating efficient electron transfer due to a high surface area.

The CPs-WS_2 nanocomposite-modified glassy carbon electrode (GCE) displayed excellent performance in detecting flutamide, with a linear concentration range spanning from 1 nM to 104 μM, a low limit of detection (LOD) at 0.74 nM, and a high sensitivity of 446.7 $\mu A\ \mu M^{-1} cm^{-2}$ in a phosphate buffer solution. Furthermore, the sensor demonstrated good recovery rates ranging from 88.47% to 95.02% in river water samples and exhibited high selectivity towards flutamide even in the presence of various inorganic ions. The GCE/CPs-WS_2 sensor exhibited impressive stability, excellent selectivity, high sensitivity, and satisfactory recovery results in complex canal/river water and artificial urine samples, highlighting its potential for determining flutamide in both environmental and biomedical applications.The Actinidia-derived CPs-WS_2 nanocomposite sensor shows considerable promise for the sensitive and selective detection of flutamide across varied environmental samples. This research underscores the importance of developing advanced sensing materials for pharmaceutical detection in water sources and proposes potential applications of carbon-based nanocomposites in electrochemical sensing. Future studies could potentially concentrate on optimizing sensor performance and exploring real-world applications in environmental monitoring.

295d Rapid and Sensitive Detection of Klebsiella Pneumoniae for UTI Diagnosis with Plasmonic Gold Nanoparticle Conjugated Aptasensor

Dr. Aniruddha Deb, Dr. Partho S.G. Pattader

Urinary tract infection (UTI), which can be caused by various pathogens, if not detected at an early stage can be fatal. It is essential to identify the specific pathogen responsible for UTI for appropriate treatment. This study describes a generic approach to the fabrication of a prototype for the non-invasive detection of a specific pathogen using a tailor-made plasmonic aptamer-gold nanoparticle (AuNP) assay. The assay is advantageous because the adsorbed specific aptamers passivate the nanoparticle surfaces and reduce and/or eliminate false-positive responses to non-target analytes. Based on the Localized Surface Plasmon Resonance (LSPR) phenomena of AuNP, a point-of-care aptasensor was designed that shows specific changes in the absorbance in the visible spectra in presence of a target pathogen for robust and fast screening of **UTI samples (Indian isolates)**. In this study, we demonstrate the specific detection of ***Klebsiella pneumoniae*** bacteria with LoD as low as **$3.4{\times}10^3$ CFU/mL**.This study shows the detection of specific pathogen using the development of a point-of-care (POC) aptasensor prototype for the quantification of the pathogen present in the UTI suspected patient along with the identification of a specific pathogen within 5 min. A generic methodology is developed that can confirm a specific bacterial presence in the clinical sample which can aid the doctor if the antibiotics should be administered. Not only that, the developed prototype may help to identify the specific drug that can be effective or ineffective due to mutations in pathogens that cause antimicrobial resistance (AMR). The developed aptasensor prototype for **K.pneumoniae** was tested against the gold standard clinical practice from a nearby Hospital (**GNRC Hospital Guwahati, INDIA**) and the results were found promising for frugal application as the costs of the aptasensor prototype and that of a single test are ~ **$ 7.47 and** ~ **$ 0.1** respectively. Our developed aptasensor has the potential to be the frugal point of care device for the detection of such pathogens and contribute significantly to primary healthcare.

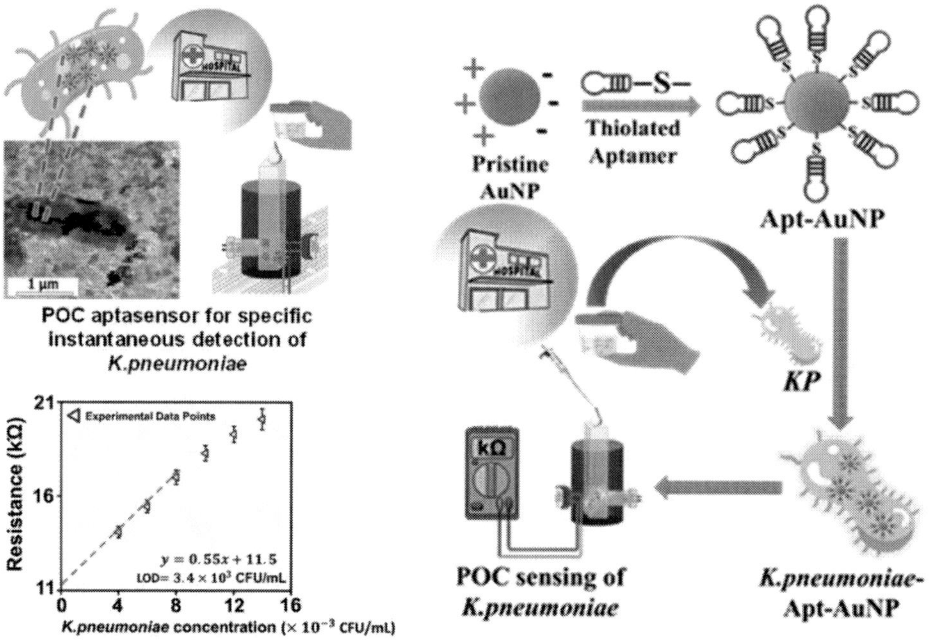

Specific Instantaneous Detection of *Klebsiella pneumoniae* for UTI Diagnosis with a Plasmonic Gold Nanoparticle Conjugated Aptasensor
ACS Applied Bio Materials **2023** *6* (8), 3309-3318
DOI: 10.1021/acsabm.3c00369

295e Quantifying Enzymatic Small Extracellular Vesicles (sEVs) for Metastatic Cancer Staging Via Magnetic Nanoporous Membrane (MNM) Activity Assay

Tiger Shi, Chenguang Zhang, Youwen Zhang, Xuemin Lu, Gaeun Kim, Sonu Kumar, Ceming Wang, Yichun Wang, Satyajyoti Senapati, Hsueh-Chia Chang

Cancer metastasis is the leading cause of death for cancer patients. Active surface proteins like membrane-type matrix metalloproteinases (MT-MMPs) and a-disintegrin and metalloproteinases (ADAMs) have been identified to be highly expressed in later stages of disease. Additionally, increased metalloproteinase (MP) expression has been observed prior to tumor metastasis to remodel distal extracellular matrices (ECM) readying the site for invasion and proliferation. Small extracellular vesicles (sEVs) – ranging from 20-200 nm – have been shown to display parent cell membrane enzymes on their lipid bilayer surface. Affinity assays offer little metastatic information captured by the activity level of the enzyme. Alternatively, existing bulk plasma activity assays for these enzymes lack sensitivity and are prone to interfering agents found in plasma.

In this work, we overcome these limitations with a Magnetic Nanoporous Membrane (MNM) diagnostic technology which uses superparamagnetic beads to rapidly isolate and enrich specific tumor cell-originated sEVs. A rapid sample incubation (30-minutes) is performed using 0.22-um filtered plasma with antibody-functionalized nanobeads. The bound sEVs are trapped by the strong field gradient present at the sharp edges of the MNM pore edge at >99% capture. A controlled wash is applied to remove non-specifically bound entities allowing for direct characterization from minimally pretreated blood plasma. The activity assay utilizes molecular beacons synthesized from enzyme-specific substrates that fluoresce upon enzymatic digestion. The platform is capable of running up to 12 wells simultaneously and can be read-out on a custom-made 96-well plate using a standard laboratory fluorescence plate reader. This assay has an overall turn-around time of two to three hours.

We establish the localization of the majority of ADAM10 enzymatic activity from cancer cells to the surface of sEVs. ADAM10+ sEV activity normalized to sEV concentration is indicative of metastatic potential in prostate cancer cell lines. Interestingly, we find that lysed sEV sample produces significantly less activity signal, suggesting that the activity of ADAM10 is conferred by its specific conformation on the sEV lipid bilayer. ADAM10+ sEV activity hence produces significant separation as a general metastatic staging biomarker in mCRC patients. The findings recapitulate existing understandings of ADAM10 in cancer metastasis and focus the field towards the sEV-bound partition. Work on the MNM well-plate platform is ongoing with significant promise in improving sensitivity down by one to two orders of magnitude compared to the bulk assay (sub-femtomolar). Our group is currently working on selective pin-down of cancer-specific EGFR+ sEVs in a larger clinical cohort of mCRC patients.

295f Comprehensive Analysis Enabled By Multi-Bore Carbon Fiber Microelectrodes: Simultaneous and Ultra-Fast Detection of Neurotransmitters and Toxic Metals

Pavithra Pathirathna, Noel Manring, Miriam Strini, Gene Koifman

Despite the significant growth in the neuroprotective drug market, there is a rising number of diagnosed neurological disorders worldwide; thus, understanding the multifactorial etiology of these conditions is crucial. While extensive research has focused on aging's relationship with disease onset, other factors such as environmental influences, particularly toxic metals, remain understudied. Additionally, there's a lack of experimental data on co-transmission's regulation and functional effects, despite its presumed role in neuronal survival control. Consequently, there's a pressing need for a tool capable of real-time, simultaneous multi-analyte detection to advance research in these areas. In our study, we fabricate double-bore and four-bore carbon fiber microelectrodes capable of concurrently detecting neurotransmitters and toxic metals using fast-scan cyclic voltammetry with a temporal resolution of 100 ms. We characterize these multi-bore electrodes with neurotransmitters like dopamine, serotonin, and ascorbic acid, as well as toxic metal ions such as copper, cadmium, and arsenic in tris buffer, simulating artificial cerebellum fluid. The nanometer-scale gaps between individual electrodes render the assembly compact. Remarkably, these electrodes withstand electrical noise when cycled with different waveforms at rapid scan rates. Furthermore, the surface of each individual electrode is functionalized with various surface modification agents to optimize performance. Calibration curves are constructed to define analytical parameters such as linear range and limit of detection. Our findings reveal higher sensitivity of multi-bore electrodes compared to single-bore electrodes, a promising step towards simultaneous detection. Preliminary data suggest translatability to in vivo detection with excellent selectivity and sensitivity, facilitating a deeper understanding of neurological disorder etiology.

295a Monitoring Electropolymerization Polymer Thickness with Various Monomers for Optimized Biosensor Design

Grace Dykstra, Yixin Liu

Molecularly imprinted polymer-based biosensors (MIPs) serve as synthetic substitutes for conventional enzyme and antibody-based sensing platforms. Using cyclic voltammetric electropolymerization, these biosensors can be directly fabricated onto transducer surfaces, simplifying the sensor design process. Despite their practicality, optimizing these sensors typically relies on trial-and-error univarient methods without a thorough understanding of the underlying polymerization principles. This research investigates the use of electropolymerization (to make eMIPs) to precisely control the thickness of eMIPs. Leveraging insights from non-imprinted polymers, the study explores how varying monomers impact polymer thickness. Through control and characterization of polymer thickness using quartz crystal microbalance alongside electrochemical measurements, the goal is to gain insight into eMIP design, and ultimately improve biosensor performance.

295h Improving Electrochemical Hybridization Assays with Restriction Enzymes

Xingcheng Zhou, Marjon Zamani, Ariel Furst

Nucleic acids in blood are early indicators of disease that could be detected by point-of-care biosensors if sufficiently sensitive and facile sensors existed. Electrochemical hybridization assays are sensitive and specific but are limited to very short nucleic acids. We have developed a restriction enzyme-assisted electrochemical hybridization (REH) assay for improved nucleic acid detection. By incorporating target-specific restriction enzymes, we detect long nucleic acids, with performance dependent on the location of the cut site relative to the electrode surface. Thus, we have further established guidelines for REH design to serve as a generalizable platform for robust electrochemical detection of long nucleic acids.

295i Health Monitoring with Breath Sensors and Devices

Andreas Guentner

Breath sensors can revolutionize medical diagnostics by on-demand detection and monitoring of health parameters in a non-invasive and personalized fashion. Despite extensive research for more than two decades, however, only few breath sensors have been translated into clinical practice.[1] While nanotechnology offers strategies to design transducers (e.g., plasmonic or chemoresistive nanoparticles) that can quantify critical compounds at sufficiently low concentrations, selectivity and reversibility remain contradictory constraints. Here, I will discuss our recent advances on the design of nanostructured breath sensors that monitor biomarkers continuously at high accuracy. This has been enabled by optimizing receptors and transducers for selective and reversible interaction for various molecule types. To demonstrate the immediate practical impact, I will present continuous sensing of breath acetone[2,3] in health and diabetes[4] that is currently evaluated in clinical trials.

[1] Güntner, A. T.; Abegg, S.; Konigstein, K.; Gerber, P. A.; Schmidt-Trucksass, A.; Pratsinis, S. E., *ACS Sens* **2019,** *4* (2), 268-280.
[2] Güntner, A. T.; Sievi, N. A.; Theodore, S. J.; Gulich, T.; Kohler, M.; Pratsinis, S. E., *Anal. Chem.* **2017,** *89* (19), 10578-10584.
[3] Weber, I. C.; Derron, N.; Königstein, K.; Gerber, P. A.; Güntner, A. T.; Pratsinis, S. E., *Small Science* **2021,** *1* (4), 2100004.
[4] Güntner, A. T.; Weber, I. C.; Schon, S.; Pratsinis, S. E.; Gerber, P. A., *Sens. Actuators B* **2022,** *367*, 132182.

295k Smart, Automatic Real-Time Water Quality Monitoring System

Swapnil Meshram, Jay Herekar, Shankar Ramachandrana, Anwesh Mangal, Perumalla Koteswara Rao, Sibi Raj B Pillai, Maryam Shojaei Baghini, Rajdip Bandyopadhyaya

Many waterbodies are experiencing escalating pollution levels as a result of waste mismanagement, expansion of industrial areas, and a lack of social awareness. The pollutants coming from small streams merge with the river, resulting in widespread contamination of many large water bodies. Therefore, tracking water quality has become increasingly essential in the present day to pinpoint and control the major sources of water pollutants.

We have developed a completely in-house manufactured water quality monitoring system, which can operate without human intervention by measuring eight key physical parameters and transmitting the data to a web interface and an Android application. The system consists of a filter and sampling ports to draw water from the waterbody with the help of a 3D-printed peristaltic pump. It also includes two 3D-printed autosamplers, electrochemical and optical sensors, microcontrollers, LoRaWAN (a low-range wireless data transmission module), and an ESP8266 receiver module.

The initiation of the system is automated through the integration of a relay and real-time clock module, facilitated by an Arduino microcontroller. After this, the MSP430 microcontroller regulates the operation of the drivers for the peristaltic pump and two autosamplers. These components work together to draw water from the water body through sampling ports, after filtering out large impurities using a filter. The mobile part of the autosampler moves angularly and vertically with precision for washing, calibration and testing of sensor probes. The sensors then measure water quality parameters by immersing them in collected water within sampling containers. The MSP430 microcontroller collects measured data, which is linked to a LoRaWAN module to transmit data to the receiver ESP8266. Subsequently, the receiver sends the collected data to the web interface and an Android application utilizing the Wi-Fi network.

The eight physical parameters are measured in the following ranges, including temperature (18-45°C), pH (1.3-12), total dissolved solids [5-5000 ppm (parts per million)], salinity [0-3.5 ppt (parts per thousand)], conductivity (8-7500 μS/cm), turbidity [10 to 8000 NTU (nephelometric turbidity unit)], dissolved oxygen (0.4-8 mg/l) and oxidation-reduction potential (-150 to +700 mV) by the system. This measurement is to deliver a more accurate assessment of water quality in different aquatic environments. Some sensors require less time for measurement, while others take more, due to chemical reactions occurring during sensing: turbidity sensors 2 sec, temperature sensors 10 sec, TDS, salinity, and conductivity sensors require 30 sec, pH sensors require 300 sec and dissolved oxygen and ORP sensors require 400 sec.

The system has exhibited excellent performance during testing, measuring eight parameters with good accuracy (mostly less than ±2, and rest less than ±5 percentage error) with ultrapure water, tap water, pond water and lake water in the laboratory, as well as in the field testing with swimming pool water and that in a water canal. In very few cases, turbidity shows a higher percentage error due to measurement at a very low level of turbidity. ORP also sometimes shows a higher percentage error, due to incomplete oxidation and reduction reactions, in order to reduce the overall process time of water quality monitoring to about 26 min in a single cycle. Such measurement and monitoring cycles can be repeated multiple times in a day and over many days, completely autonomously, for a possible time series analysis of water quality parameters.

In addition, the system has many advantages, reduced contact between probes and contaminated water by drawing water from the testing container via the peristaltic pump, avoiding water storage, probe washing before testing new water sample, autosampler operation, customized PCBs and wireless data transmission to an Android app. These features also introduce novelty to the overall process. The system has a compact size of 19.5 cm x 12.5 cm x 21.5 cm, weight (17.5 Kg) and an approximate manufacturing price of $836.

The system can be deployed at different locations at various waterbodies. This can help to identify zones such as low-polluted, moderately polluted, and highly polluted. The system can be deployed in fishing areas and swimming pools to prevent situations like low dissolved oxygen and high turbidity.

Another system has been developed for monitoring both metal and non-metal ions, such as arsenic, chromium, fluoride, and iron by a colorimetric technique using respective reagents. Both systems have similar features like automatic operation, LoRa wireless data transmission, and Android application access, offering significant advantages in ensuring water safety and environmental health.

2951 A Novel, Ultra-Fast Electrochemical Detection of Neurotoxic Cadmium and Arsenic Using Carbon Fiber Microelectrodes and Fast-Scan Cyclic Voltammetry

Noel Manring, Miriam Strini, Gene Koifman, Pavithra Pathirathna

Neurotoxic agents like Cd(II) and As(III) pose a significant global health risk, especially through contamination of drinking water. Prolonged exposure to these toxins via water, food, and air can lead to various health complications, including cardiovascular diseases, neurological disorders, cancer, and even death. Hence, there's a critical need for highly sensitive and rapid detection methods to identify these heavy metals in biological systems. Traditional metal detection methods typically involve extracting samples from the body, which alters metal speciation, a key factor in determining toxicity. Moreover, these methods aren't suitable for real-time measurements in living organisms. While electrochemists have devised detection methods for neurotoxic agents, they often lack specificity regarding metal speciation and are primarily applicable to environmental samples, rendering them unsuitable for in vivo analysis. In our study, we've optimized fast-scan cyclic voltammetry (FSCV) to detect Cd(II) and As(III) using carbon fiber microelectrodes (CFMs) with a temporal resolution of 100 ms. We employed gold nanoparticles (Au NPs)-modified CFMs for Cd(II) detection and bare CFMs for As(III) detection. Characterization of both sensors was conducted in a tris buffer mimicking artificial cerebellum fluid. Calibration curves were established for both analytes to determine the limit of detection, linear range, and sensitivity. Additionally, comprehensive selectivity tests were performed in the presence of other interfering ions to assess sensor specificity. Furthermore, we conducted an in-depth pH study, particularly for As(III), given its complex aqueous chemistry dependent on medium pH. Our research demonstrates that Au NPs-modified CFMs can detect ultra-low concentrations of Cd(II) in artificial urine samples. Notably, this study marks the first instance of Cd(II) and As(III) detection under ambient conditions using FSCV and CFMs at ultra-fast temporal resolution. Our preliminary findings suggest promising prospects for utilizing these sensors in future in vivo applications, particularly within the brain.

295m Inoses: Intelligent Nature-Inspired Olfactory Sensors Engineered to Sniff

Haritosh Patel, Anna Shneidman, Joanna Aizenberg

The need to acquire real-time information about the air we breathe has been brought into the spotlight through recent events, such as wildfires, the COVID pandemic, hazardous spills, and increasingly stringent emissions regulations. Standard methods to determine the composition of gaseous samples typically rely on bulky and expensive spectroscopic equipment, necessitating the transportation of potentially hazardous materials to the monitoring site. This increases risk and contributes to a high environmental footprint. Despite significant advances, current portable sensors are still limited in their ability to accurately analyze a diverse array of volatile compounds and mixtures. Dogs are still employed in many scenarios (such as detection of explosives and contraband, as well as, more recently, early disease detection), highlighting the impressive capabilities of the natural olfactory system. Meanwhile, sensors for volatile organic compounds (VOCs) and other hazardous air pollutants (HAPs) are often reported to be unreliable, difficult to calibrate, suffer from drift, and are highly susceptible to humidity and temperature changes.

In order to address the need across various industries for low-cost, accurate and real-time solutions for identifying volatile mixtures, we have developed intelligent Nature-inspired Olfactory Sensors Engineered to Sniff (iNOSES). Distinctive to our approach, the device implements biologically inspired sniffing: it generates and self-optimizes adaptive patterns of inhale-exhale sequences, which capture non-equilibrium mass transport phenomena experienced by each compound or mixture due to differences in their physicochemical properties, mimicking critical natural odor discrimination mechanisms. The presence of analytes is recorded in the form of time-resolved sensor output that is featurized to enable machine learning (ML) techniques for analyte detection and classification. Thus, discrimination of analytes is achieved *via* the interactions between the "olfactory system" (our sniffing device) and the "brain" (represented by ML models). We demonstrated that a sensor of this type, which incorporates non-equilibrium mass-transport dynamics, temporal data collection, and ML modeling, substantially enhances the detection power of any artificial nose, including – as is in our case – a single, mesoporous one-dimensional photonic crystal, which we demonstrated could distinguish several volatile polar and non-polar compounds as well as their binary mixtures, and could predict physical properties of known and unknown compounds. The approach is hardware-agnostic and can be applied to practically any volatile sensor, allowing us to broaden the repertoire of detectable gasses. The vision of iNOSES is to enable individual consumers, government agencies such as the US EPA, residential and non-residential building entities, and more to have access to accurate, affordable, real-time volatile sensors to make reliable measurements and predictions of indoor and outdoor air composition. This is a critical step to improving human and ecosystem health as well as defining and standardizing emissions regulations.

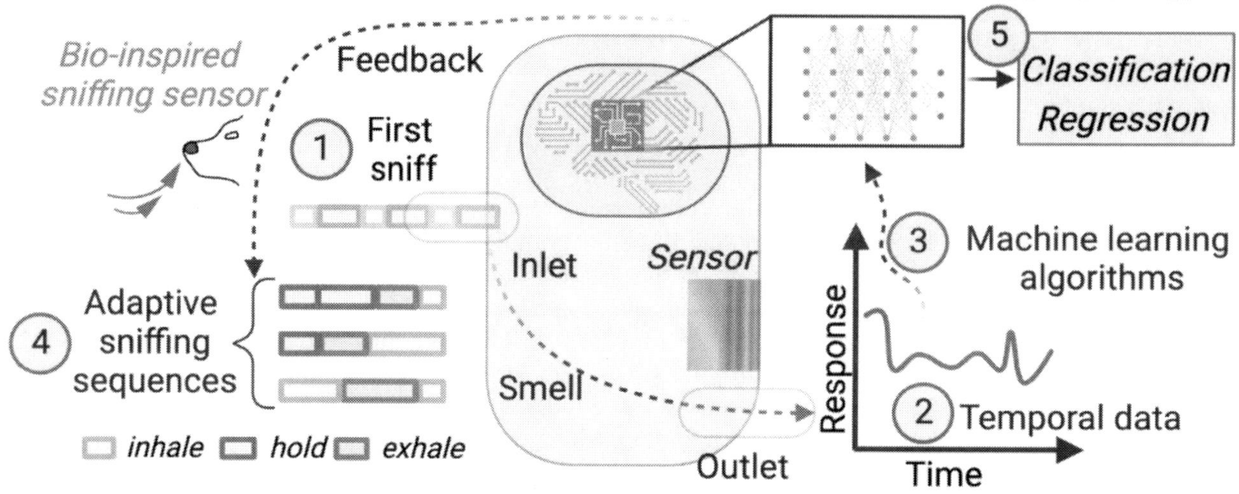
Bio-inspired
sniffing sensor
Feedback
1 First sniff
Inlet
Sensor
4 Adaptive sniffing sequences
Smell
Outlet
inhale
hold
exhale
5 Classification
Regression
3 Machine learning algorithms
Response
2 Temporal data
Time

295n Multiplex Electrochemistry: Translation to Quantitative Microarray for Medicine

Joydip Dey, Bhavana Sunil, Thomas Hope, Ravi Saraf

Immobilization of DNA oligomers on Au has been pervasively applied to a broad range studies. Primary motivation has been the possibility of electrochemical detection to quantify DNA binding to microarray for early diagnosis of cancer, cardiovascular and neurological diseases. However, the tantalizing fundamental results of a "perfect switch" between ssDNA versus dsDNA signal for the electrochemical beacon (EcB) method discovered two decades ago (Fig. 1(a)) has not been translated. We studied EcB by local electrochemistry on 10 to 100 μm spot by an instrument developed by Vajra Instruments called SEED.[1] We observe a distribution of conformation making the switch imperfect (Fig. 1(b)). We observed that,

- before, MCH backfilling the MB on ssDNA has total freedom to undergo redox but contrary to "perfect switch" mechanism redox appears after (Fig. 2); and
- the peaks (R_{max}) for oxidation and reduction measured by SEED on the same chip shows that the decrease in the signal on binding depends on the length of the probe (Fig. 3).

Based on these observations we have developed a strategy to increase the Limit of Quantification (LOQ) from 1 picomolar in previous studies to 10 attomolar that is significantly better than qPCR for short chain nucleic acids such as, microRNA (Fig. 4). We will describe the modified mechanism leading to the strategy that improved the performance by orders of magnitude.

1. Tevatia, R.; Prasad, A.; Saraf, R. F. Electrochemical Characteristics of a DNA Modified Electrode as a Function of Percent Binding. *Analytical Chemistry* **2019**, *91*, 10501-10508.

360a Enhanced Sensitivity in Cortisol Detection Using Beta Cyclodextrin Biosensor: Integrating Automated Flow Cell System with EIS

Ferdows Sajedi, Jeffrey Halpern

This study aims to assess the efficacy of a βCD biosensor in non-redox mediated electrochemical impedance spectroscopy (EIS) for measuring cortisol levels in PBS and urine. Cortisol presents a unique challenge in electrochemical sensing, being electrically neutral at physiological pH and lacking electrochemical activity, necessitating innovative strategies. Because cortisol plays a crucial role in controlling many important body functions like immunity, metabolism, and heart health, there is a pressing need for accurate detection methods. We improve our methodology based on previous successful experiments conducted with static cells [1] by converting it to automated flow cell system.

In this study, we focus on measuring different variables of EIS across various frequencies to optimize sensitivity of the sensor. Surface modification involves grafting diazonium salt onto a glassy carbon electrode, followed by carboxylic group conjugation using EDC/NHS chemistry. Further modification with polypropylene glycol (PPG) precedes cyclodextrin loading onto the surface. Characterization of the surface is carried out through EIS to characterize the polymer-modified surface state in analyte detection.

Our results demonstrate remarkable sensitivity to cortisol, even at low concentrations, representing a substantial improvement over static cell method [1]. Additionally, we are testing the stability, aiming for long-term stable automated system for analyte detection. These finding highlight the clinical potential of our biosensor before we transition into monitoring cortisol levels for various health conditions.

Research supported by New Hampshire- INBRE through an Institutional Development Award (IDeA), P20GM103506, from the National Institute of General Medical Sciences of the NIH. Support was also provided by a National Science Foundation EPSCoR award (#2119237), BIO-SENS.

[1] Z. Panahi, T. Ren, and J. M. Halpern, "Nanostructured Cyclodextrin-Mediated Surface for Capacitive Determination of Cortisol in Multiple Biofluids," *ACS Appl. Mater. Interfaces*, vol. 14, no. 37, pp. 42374–42387, Sep. 2022, doi: 10.1021/acsami.2c07701.

360b Anodically Grown Pt(II) Oxide Microelectrode/Nanoelectrode pH Sensor

Abdullah Bahdad, Kevin C. Leonard

Platinum electrode availability in electrochemical laboratories makes it attractive as a potentiometric solid-state pH sensor for aqueous solutions. However, challenges have been reported in fabricating anodic-grown platinum oxide potentiometric pH sensors. In this work, cyclic voltammetry in 0.5 M H_2SO_4 and 1 M NaOH was conducted to assess the effect of acidic and basic solutions on the anodic-grown platinum (II) oxide for designing Pt/Pt(II) oxide pH sensor. Chronoamperometry was also employed to grow a platinum (II) oxide layer on an ultramicroelectrode in both solutions, followed by measuring their potentiometric response in various pH solutions. The platinum (II) oxide manufactured in 1 M NaOH showed a close Nernstian response (~ -60 mV/pH), and was insensitive to dissolved oxygen, cation and anion identities, and molar ionic strength. When the platinum (II) oxide was grown on a platinum nanoelectrode, the response time was ~ 2 seconds for an approximately twelve-fold pH step change. These findings highlight the importance of the solution pH used during platinum oxidation on anodic platinum (II) oxide properties in sensing solution pH. The straightforward approach used in this work for fabricating Pt/PtO microelectrode/nanoelectrode pH sensors by direct oxidation can ease real-time pH measurement in complex aqueous environments.

360c A Comprehensive Strategy for Whole-Cell Electrochemical Biosensor Design: From Sensor Biomining to Deployable Devices

Alec Agee, Ariel Furst

The ability to detect contaminants, soil nutrients, and other relevant molecules in the environment is crucial to protect human and environmental health. However, established measurement techniques for these analytes are typically expensive, laborious and slow. Alternative sensing technologies, namely whole-cell electrochemical biosensors, are increasingly popular alternatives. These technologies use microbes to detect analytes, harnessing the array of naturally-evolved sensor systems in living cells. These microbes then interface with electrodes, producing electrical signals that can be interpreted by digital devices. These sensors boast high specificity and low cost, and we have integrated the development of these sensors into a comprehensive and generalizable workflow. We have streamlined the receptor biomining process through the development of a bioinformatic tool and guiding principles for selecting and expressing sensor proteins. We have additionally developed a quantitative model of gene circuit performance in the platform sensor strain Shewanella oneidensis, enabling a priori design of electroactive gene circuits with an arbitrary receptor protein. Finally, we have invented a novel device configuration which reduces cost by an order of magnitude while eliminating requirements for specialized equipment. We have used this workflow to develop biosensors for a suite of soil nutrients and environmental contaminants, demonstrating the broad applicability of our strategy.

360d Four-Bore Carbon Fiber Microelectrodes for the Simultaneous Detection of Heavy Metals and Neurotransmitters Via Fast Scan Cyclic Voltammetry

Noel Manring, Miriam Strini, Gene Koifman, Pavithra Pathirathna

Neurodegenerative diseases (NDDs) pose an increasing burden on global health, with conditions such as Alzheimer's disease, Parkinson's disease, and amyotrophic lateral sclerosis affecting more than 50 million people globally. The etiology of these conditions is complex and multifactorial making the development of efficient therapeutics to slow down or stop progression challenging. Although the mechanism of these diseases is not well understood, the identification of abnormal levels of neurotransmitters has been linked to patients suffering with these disorders. These abnormal levels are correlated to the loss of motor functions, dementia, and overall cognitive incline that is associated with these diseases since neurotransmitters are responsible for neural communications. Additionally, epidemiological studies have implicated heavy metals in increased NDD risk, with chronic exposure posing a significant threat to neurological health. Understanding the intricate network of factors driving NDDs is critical for the development of effective therapeutics. Electrochemists have extensively used carbon-fiber microelectrodes (CFMs) to better understand the pathology of NDDs over the last few decades. However, most studies analyze one neurotransmitter or metal at a time with a single CFM, thus, limiting the real-time information about interactions between neurotransmitters and other factors. Recently, double-bore CFMs have been reported to study the interaction of two analytes, revealing more crucial information on NDDs; however, it would be beneficial to have an electrode with even more sensing components. In this study, we fabricated a four-bore CFM with four separate sensing components to analyze four analytes using their analyte-specific waveforms. We characterized our sensor with dopamine, serotonin, ascorbic acid, Cu^{2+}, Cd^{2+}, As^{3+}, and Cr^{3+} via fast-scan cyclic voltammetry (FSCV). We acquired FSCV signals simultaneously in analyte mixtures prepared with four analytes in tris buffer. We constructed calibration curves showcasing the stability of our multi-bore electrode and ultra-fast scan rates. To the best of our knowledge, this is the first study reporting use of four-bore CFM to detect four analytes simultaneously without altering any electrochemical parameters at the temporal resolution of 100 ms; thus, showcasing an excellent possibility to use this sensor in vivo for real-time detection of multiple neurotransmitters and other toxic metal ions.

360e Aptamer-Based Biosensor for Dimethylsilanediol Detection

Niloofar Zolfigol, Zehui Xia, Paul Nahass, Baikun Li, Yu Lei

Dimethylsilanediol (DMSD) emerges as a concerning contaminant in drinking water on the Space Station, primarily originating from the breakdown of silicon-based compounds in personal care products. Its presence in water collected from humidity condensation poses health risks to astronauts, and thus there is an urgent need to develop a sensor or tool for detecting DMSD accurately. To address this challenge, we are developing aptasensors for detecting DMSD in water samples.

Aptamers, single-stranded oligonucleotides with precise binding capabilities, offer a versatile means of identifying a wide range of contaminants. Selective aptamers are screened through a process named Systematic Evolution of Ligands by Exponential Enrichment (SELEX). In this study, we will first identify DMSD aptamers through Capture-SELEX, a modified SELEX technique that involves immobilization of the DNA library onto magnetic beads to enhance screening efficiency. Then we will employ advanced technologies to characterize and validate the aptamers identified. Finally, we will construct an aptamer-enabled biosensor for DMSD detection. By harnessing the unique binding properties of aptamers, the detection of DMSD in water samples with high sensing performance will be achieved.

360f Electro-Spun $ZnMoO_4/ZnCo_2O_4$ Nanofibers-Based Gas Sensor for Ppb-Level n-Butanol Detection

Asky Fungura, Yixin Liu

1-Butanol, a volatile organic compound (VOC) is a common solvent used in industry that has the potential to find its way into residential areas through its usage in paints, pesticides, and cosmetics. Effects of inhaling high concentrations of 1-butanol include dizziness, headache, drowsiness, and dermatitis. Metal oxide gas sensors have been chosen as one of the economical, fast, and reliable solutions for Volatile organic compound (VOC) gas testing. The metal oxide chosen was $ZnCo_2O_4$ spinel and was electro-spun into nanofibers,1D nanomaterials used to enhance sensitivity and response. More research is being done into the p-type, to match their n-type metal oxide counterparts and spinels are a sub-branch within some of the p-type metal oxides. Low response to gases is the major problem within p-type gas sensors and p-n heterojunctions were used to improve by adding $ZnMoO_4$ to the p-type spinel. Characterization of the morphology was done by Scanning an electron microscope which showed nanofibers and x-ray diffraction confirmed the presence of the spinel in all the material. High Resolution-XPS showed that mixing $ZnMoO_4$ and $ZnCo_2O_4$ improved surface science to enhance 1-butanol sensing. Gas testing using air as the diluting medium showed the best selectivity and a response of 4.94 to 10 ppm was achieved. The lowest concentration tested was 250 ppb with a response of 0.124. Using p-type metal oxide sensors provides diversity to the material used for 1-butanol gas sensing. Ppb gas sensing allows fast detection of tiny amounts of 1butanol in the environment below the human olfaction limit of around 0.8 ppm.

565d The importance of model systems in biology and materials science: A measurement perspective

Matthew Ryen Lockett

Models provide an experimentally tractable means of studying complex systems whose interconnected components can result in non-additive and non-obvious responses. Guided by measurements—some new, some established, and others reimagined forms of the tried-and[1]true—we develop models to systematically dissect aspects of complex phenomena that impact our everyday lives in the areas of solar fuels and drug discovery. In the area of solar fuels, we quantify the effect of surface chemistry on the performance of silicon- and carbon-based photoelectrodes. The electronic properties of (bulk) silicon, diamond, and graphite are well characterized. The surface of these materials has proven challenging to study due to the low concentration of surface states and chemically appended molecules (a limit of detection problem) and their reactivity in operando and under ambient conditions (a stability problem). Using silicon as a model semiconductor surface, we evaluate: 1) how introducing dipoles on the surface alters the favorability and kinetics of electron transfer from the surface to appended catalysts. 2) The effect of lateral interactions on the electrochemical reversibility of surface-appended catalysts. 3) The effect of electric fields on the ordering and activity of surface-appended catalysts. We apply these findings to amorphous carbon films, representing a disordered (and more challenging to characterize) chemical environment. In the area of drug discovery, we focus on developing physiologically relevant tissues for in vitro screening applications. Our laboratory develops devices and methods to evaluate the role of oxygen on cellular behavior in tissue-like environments. By treating oxygen as a reagent, we specifically focus on how traditionally relied-upon culture methods (e.g., atmospheric or 20% O2 levels) overlook important cellular responses occurring at physiologically relevant oxygen tensions. Some of our recent findings show oxygen should be accounted for when predicting (1) drug sensitivity and resistance in tumor models, (2) the potential toxicity of environmental contaminants, and (3) the metabolic activity and competency of healthy hepatocytes in liver models

AUTHOR INDEX